9/10/71

buckeye · ewbank · ginther · midwest publications

a cloudburst of math lab experiments

Midwest Publications
P.O. Box 307
Birmingham, Michigan
48012

SBN 910974-33-0

a continuum approach to math lab experiments

TEACHER'S MANUAL

ADVICE ON ORGANIZING A MATH LAB

Math lab - - What is it?

A math lab may be viewed as a mode of classroom management in which students, working individually or in small groups, do mathematics experiments. Each of these experiments presents the student with activities (usually with physical apparatus) which have mathematical implications. The books comprising the *Cloudburst* series contain approximately one thousand such experiments.

To the authors, the term "math lab" refers more to a system or method of operation than to a particular place in the school building set aside for mathematics experiments. In fact, it may not be the best thing to have an actual "lab room," because of scheduling problems, etc., but any teacher can use the math lab approach in his classroom, and at minimal expense.

Some advantages of the math lab approach are: the possibility of enhancing the student's motivation to learn mathematics, the presence of a natural learning sequence of "concrete-to-abstract," and the opportunity to discover real-world applications of mathematics.

When math lab?

Those experienced with running a math lab of this type advise: "Begin small, expand gradually!" Perhaps initially use the experiments for enrichment only, for those who have finished their other math work. Once this use of the experiments is going smoothly, perhaps one day per week could become math lab day for the entire class. As teacher and students become accustomed to the math lab, a greater proportion of time could be devoted to math lab, if desired. A few teachers use the lab approach for the total math program, using standard textbooks only as reference sources. We repeat our caution, however: "Begin *small,* expand *gradually*!" The math lab does require some planning and it may take students awhile to get used to working in a freer situation.

Some curriculum considerations

To facilitate the coordinating of concepts included in math lab with those in the typical textbook setting, each *Cloudburst* experiment is coded to indicate the concept it reinforces. Using the explanation of the coding system (later in this Preface), a teacher can locate those experiments that deal with a specific topic. The teacher then can prescribe pertinent experiments for his various students.

Teams

Those with experience have found that there are benefits in having students work in pairs on the experiments. This permits student discussion and mutual assistance as they collaborate on an experiment. Some teachers decide what students will work together, while others permit the students to choose their own partners. In ordinary situations, the latter practice may lead to happier students!

Assigning and evaluating student work

The teacher can prescribe experiments, or allow students to choose their own, perhaps from a selection suggested by the teacher.

One reporting system has each member of a lab team do a write-up of a finished experiment (a sample write-up form appears on p. vii) - - what he did, how he did it, and what his findings were. These write-ups can then be deposited in that individ ual's file folder. The teacher then can assign a "+" or a "0" to indicate "satisfactory" or "unsatisfactory" and leave the write-up in the folder for the student.

The following set of questions provides a way of orally evaluating students' work on experiments. No particular sequence of questions is necessary, but rather the teacher selects those most appropriate.

A. Appropriateness of the experiment:

1. Why did you choose this particular experiment? Was this a good choice?
2. Could you tell what was happening all the time? Was there some part of the experiment that you did not understand? If so, how did you straighten yourself out?
3. Was this experiment hard, or easy for you? What made it hard or easy?
4. Did you want to keep on with the experiment, or did you have to force yourself to finish it?

B. Values gained:

1. Tell me a few high points of the experiment.
2. Which part of the experiment was most important? Why?
3. What do you think the person who wrote this experiment was trying to say to you?
4. What new ideas did you learn from doing this experiment?
5. What do you think you could use this information for, besides what it was used for here?
6. If I made a change [specify a particular change appropriate to the experiment] what would the result be?
7. What new questions or ideas does this experiment suggest to you?

Of course, it is possible to use a combination of written and oral reporting, and this may be preferable in a typical situation with all except the very young students.

A procedure for records-keeping involves the use of a grid chart, either a large one posted for all to see, or a small one kept privately by the teacher. The chart would show students' names and experiment numbers, e.g.:

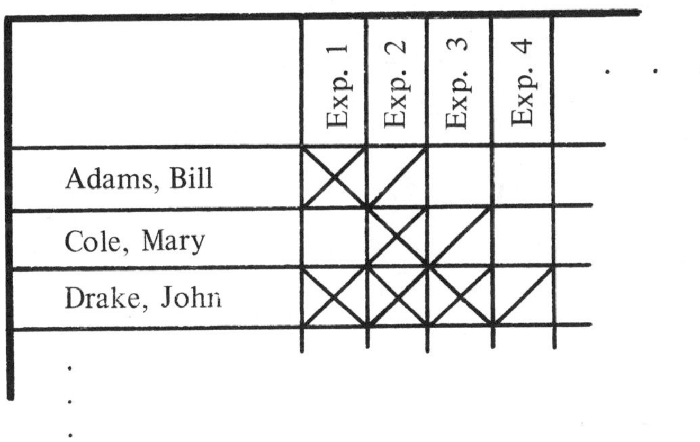

The single diagonal in a chart cell indicates that the student is currently working on that experiment. When the experiment is satisfactorily finished, the "X" mark is completed. Such a chart gives the teacher a quick visual record of class progress. When a student has "X'ed" an experiment, he then is an "expert" on that experiment, and anyone who later does that experiment is urged to consult such an "expert" if he needs help - - ask the teacher as the last resort! (This is one way of conserving the teacher's time).

Apparatus

Each experiment indicates what apparatus is needed. Apparatus can be stored on open shelves, in a cabinet, or in cardboard boxes (shoe boxes work well).

Students can be trained in good "housekeeping" habits in learning to put things back where they found them, after use.

It will be found that much - - nearly all - - of the apparatus can be improvised by using materials readily available around the house, from a hardware store, etc. Often, students can make the apparatus (which can be preferable educationally to purchasing expensive ready-made equipment). This math lab can be "launched" on practically a "zero" budget. Also, since each lab team can work on a different experiment, only one set of equipment is needed for any experiment.

The experiments can be transferred to card form, then each card can be covered with transparent contact paper. Use of such cards permits convenient display of the experiments, since the cards can be inserted into envelopes glued onto poster board.

What about discipline problems?

In a relatively "free" classroom situation such as the math lab, the teacher might expect to face bedlam. In practice, however, teachers we know who have tried this math lab system invariably find that discipline is perhaps less of a problem than usual, since students are actively involved in the experiments.

One system for "ensuring" reasonable conduct is to have clearly-stated citizenship rules for math lab, then mark infractions on a simple grid chart.

And finally . . .

If the teacher finds these experiments useful in motivating his students (and we feel he will!), he probably will soon be writing additional experiments of his own. And, very likely, *students* will begin writing new experiments. Good ideas are readily available, and the experiments are not difficult to write.

HOW THE CODING SYSTEM WORKS:

Each experiment in each *Cloudburst* book has a four-digit code number, assigned as in the following example:

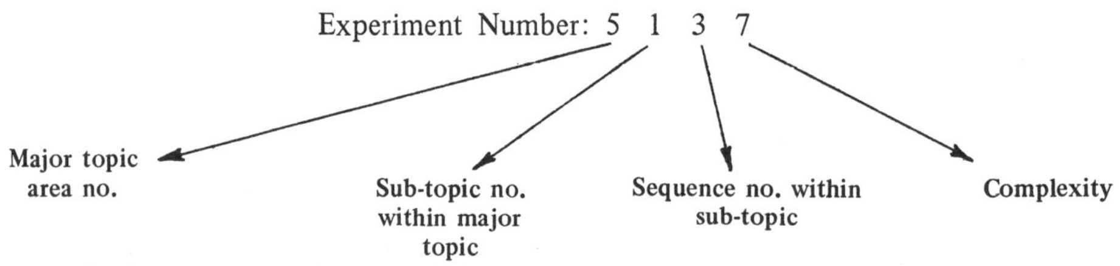

Therefore, Experiment No. 5137 is an Application (5000) experiment on length (100). It is the 3rd experiment on length with a 7 complexity. [1 is the lowest complexity and 9 is the highest complexity].

Note that this coding system is used to systematize the *entire* set, approximately of one thousand experiments which comprise the four volumes of *Cloudburst* books. Hence, not all of these topics may appear in any individual book of the series.

A SAMPLE EXPERIMENT WRITE-UP FORM

Name:

Partner's name:

Experiment no. and title:

Attach all work sheets and graphs.

Discussed experiment
with teacher __________

You now are an EXPERT
on this experiment! __________

VOLUME 1

NOTE: *Experiments not listed* are experiments with student research answers. The student research answers will vary.

1111 3; 7; red = 2; maroon = 4; yellow = 5; light blue = 6; brown = 8; dark blue = 9; orange = 10.

1121 Various sets of addends that add to a given sum

1131 3) Subtraction

1211 Use it for subtraction

1331

2	12
3	8
4	6
6	4
8	3

1371 2) Six ways

1421 This is the "greatest common divisor" (GCD) table

2131 3) yellow=5; orange=10; red=2; green=3; maroon=4; light blue=6; black=7; brown=8; dark blue=9

2221 1) black=7; orange=10; red=2; green=3; maroon=4; yellow=5; light blue=6; brown=8; dark blue=9; orange=10

2231 1) 24

2241 Number of orange rods tells how many tens

2261 1)

2271 3)

2411 3) Add or subtract 1

2431 1) All cover the same amount of floor

2931 Odd

2941 1) 111 4) 444 7) 777
2) 222 5) 555 8) 888
3) 333 6) 666 9) 999

2951 6) No remainder

3121 1) 12 inches 3) 4 inches
2) 6 inches 4) 3 inches

3131 1) Each jar is half full $1 - {}^1/_2 = {}^1/_2$
2) Half quart = one pint

3141 1) 21 3) $^1/_2$
2) $^1/_2$; $^1/_2$ 4) $^{15}/_{21}$; $^1/_{21}$

3151 1) $^1/_2$ 5) $^1/_2$
2) $^1/_8$ 6) $^1/_{16}$
3) $^1/_2$ 7) $^3/_{32}$
4) $^1/_4$ 8) $^1/_8$

4322 2) 18 ; 22

5151 1) The distance from thumb end to little-finger end of a spread hand

5541 1) 1 hour 4) 12
2) 12 hours 5) 12
3) 12 times 6) 24

5711 It remains level; horizontal

5721 2) Vertical

5741 a) 90° e) 180°
b) 90° f) 180°
c) 90° g) 45°
d) 90° h) 45°

6162 2) It took 8 straws and 8 pipe cleaners

6171 1) Angle 4) Parabola forms

6211 1) 21 2) 13

VOLUME 1

6242 a) 5 b) 4

6261 Rectangle or square

6281 a) 14 b) 25

6291 Different

6292 2) Triangle – ½ area of square
3) ¼ area of square
4) $^1/_8$ area of square

6301 Circle

6331 4, 5

6411 a) MAN e) OH
b) MOM f) HA
c) MAT g) MA
d) TIM

6421 1) Yes 2) Yes

6431 1) No
2) Yes, Yes
3) Yes

6441 1) A, B, C, D, E, H, I, K, M, N, O, T, U, V, W, X, Y
2) 1 3 8
3) Yes
4) Yes

6512 Angles

7212 4) One is a line, the other is a square

7262 Parabola

7282 2)

1	6
2	24
3	54
4	96
5	150
6	216
7	294
8	384
9	486
10	600

3) $6 \bullet 1000 \bullet 1000 = 6{,}000{,}000$ square units
$6 \bullet n^2$

7312 String for figure 1 is 2 times string of figure 2

7342

1	1	1
2	4	8
3	9	27
4	16	64
5	25	125
6	36	216
7	49	343
8	64	512
9	81	729
10	100	1000

8111 a) all but ball and egg
b) all but box, prism, pyramid

8121 a) cans, bottles, balls
b) boxes, triangles, circle, square, can, bottle
c) cans, bottles

8131 2) circle, rectangle, triangle
3) rectangle and triangle
4) triangle
5) circle
6) inside the circle and outside the triangle
7) yes

8151 North pole, white
10 miles north of South pole

8311 One solution:

11	13	3
1	9	17
15	5	7

8321 One solution:

8	6	16
18	10	2
4	14	12

8331 One solution:

1	2	3	4
2	3	4	1
3	4	1	2
4	1	2	3

1	2	3	4
4	3	2	1
2	1	4	3
3	4	1	2

VOLUME 1

8341 One way:

2	1	3	2
2	1	1	4
2	2	3	1
2	4	1	1

8351 One way:

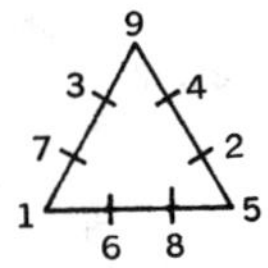

8711 Total should be even number until 5 are left

8761 1) 12 of each color
2) 24 ; 8
3) 32 of each color; 204 (counting overlapping ones)

8771 1) 10
2) 121

VOLUME 2

NOTE: *Experiments not listed* are experiments with student research answers. The student research answers will vary.

1123 Distance times washers should equal the number of washers times distance.

1143 I) Commutative property
II) Associative property

1153 Three to the left.

Any move to the right must be balanced with a equal number of moves to the left.

1163 2) 3 cents = (1) (1) (1)
4 cents = (1) (1) (1) (1)
5 cents = (1) (1) (1) (1) (1)
6 cents = (1) (1) (1) (1) (1) (1)
16 cents = (10) (1) (1) (1) (1) (1) (1)
26 cents = (10) (10) (1) (1) (1) (1) (1) (1)
36 cents = (10) (10) (10) (1) (1) (1) (1) (1) (1)
37 cents = (10) (10) (10) (1) (1) (1) (1) (1) (1) (1)
38 cents = (10) (10) (10) (1) (1) (1) (1) (1) (1) (1) (1)
48 cents = (10) (10) (10) (10) (1) (1) (1) (1) (1) (1) (1) (1)
58 cents = (10) (10) (10) (10) (10) (1) (1) (1) (1) (1) (1) (1) (1)

1164 2) $4.80

1173 2) All amounts from 1 cent through $1.65

1313 Equals multiplied by equals are equal.

Equals multiplied by unequals are unequal in the same order.

1523 Here are just a few of the answers. There are many correct answers.

$$3 = \frac{(4 \text{ X } 4) - 4}{4}$$

$$5 = \frac{(4 \text{ X } 4) + 4}{4}$$

$$6 = \frac{4 + 4}{4} + 4$$

$$7 = 4 + 4 - \frac{4}{4}$$

$$8 = 4 + 4 + (4 - 4)$$

$$9 = 4 + 4 + \frac{4}{4}$$

1533 0, 1, 2, 3, 10

All answers are one.

2123 2) 32
3) 8 sets: small circles, large circles, small squares, large squares, etc.

2144 1) Rectangle – {1, 3, 4, 6}
2) Triangle – {2, 3, 4, 5}
3) Circle – {4, 5, 6, 7}
4) Rectangle & triangle {3, 4}
5) Rectangle & circle {4, 6}
6) Rectangle, triangle, circle {4}
7) Triangle *or* circle {1, 3, 4, 6, 7}
8) Square *or* triangle *or* circle {1, 2, 3, 4, 5, 6, 7}
9) Rectangle, *not* circle {7, 2, 8}
10) Triangle, circle *not* rectangle – {4}
11) Rectangle, circle, *not* triangle – {6}

2153 1) 4
2) 8
3) 2; 16
4)

0	1
1	2
2	4
3	8
4	16
.	.
.	.
.	.
12	2^{12}
.	.
.	.
.	.
n	2^n

2333 a) VV = X
b) XX
c) LL

yes

2513 2) These are called *"prime"* numbers.

3)

1	1	1	1
2	2	2	2
3	+ 3	4	4
6	6	7	7
		14	+ 14
		28	28

6 and 28 are *"perfect"* numbers.

VOLUME 2

2514 2) 1 divides every number.

3) 1 – {1}
2 – {1, 2}
3 – {1, 3}
4 – {1, 2, 4}
5 – {1, 5}
6 – {1, 2, 3, 6}

4) Only two factors.

5) Prime numbers only have two factors.

6) Composite numbers.

2523 1) {3, 5} , {5, 7} , {11, 13} ,

2) many

3) three primes and two of them differ by only two {3, 5, 7} .

4) yes

2613

12	18	1G, 1R	3 × 2
12	20	2R	2 × 2
20	36	2R	2 × 2
24	36	2R, 1G	2 × 2 × 3
18	20	1R	2
18	36	1R, 2G	2 × 3 × 3

2623 3) One use: reducing fractions to lowest terms.

2713 1) At 5 four-rods and 4 five-rods.

2723 3) One use: finding least common denominator when adding fractions.

2813 a)  etc.

b) yes

c) 2 . . or :

d) 64 + 36 = 100 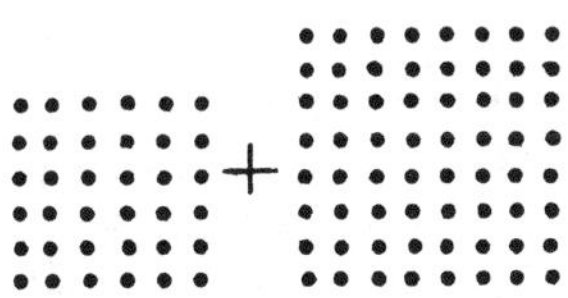

2814 square array

square numbers

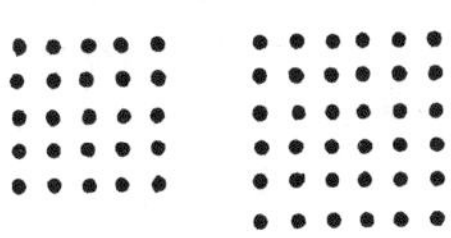

1 + 4 = 5
1 + 4 + 9 = 14
1 + 4 + 9 + 16 = 30
1 + 4 + 9 + 16 + 25 = 55

$$\frac{2n^3 + 3n^2 + n}{6}$$

2823 4)

1	1
2	4
3	9
4	16
.	.
.	.
.	.
7	49
.	.
.	.
.	.
25	625

2833 1) No. of dots = (no. of ┘'s) ^{2}
No. of new dots = 2 × (no. of ┘'s) – 1

2834 2) 4; 9; 16; 20 × 20.

3) 11 × 11

2843 1) 36 47 57 68 80

2) 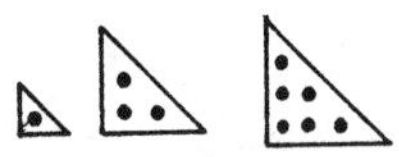 etc.

3) You get perfect squares.

2844 Triangular array, 1, 3, 6, 10, 15

1 + 3 = 4
3 + 6 = 9
6 + 10 = 16
10 + 15 = 25

squares

1 – 0 = 1
3 – 1 = 2
6 – 3 = 3
10 – 6 = 4
15 – 10 = 5

yes, the counting numbers.

2853 1) No. of dots = $\frac{\text{(No. of diagonals)} \times \text{(1 + no. of diagonals)}}{2}$

No. of new dots = no. of diagonals.

2854 2) 3; 6; 10; $\frac{20 \cdot 21}{2} = 210$

3) $\frac{12 \cdot 13}{2} = 78$

2863 2) 27

3) 64; 125; 216

VOLUME 2

3113 largest triangle – 2 times
smaller triangles – 8 times

3123 rectangle, square
largest rectangle – 2 times
smallest rectangle – 8 times
largest square – 4 times
smallest square – 16 times

3133 1) $^1/_{12}$; $^2/_{12}$, etc.

3213 1) $^4/_3$ or $^1/_9$
2) $^8/_{36}$ or $^2/_9$
3) $^2/_6$ or $^1/_3$
4) $^{10}/_{36}$ or $^5/_{18}$
5) (a) $^2/_9$ (b) $^1/_6$ (c) $^1/_4$ (d) $^7/_{36}$ (e) $^7/_{18}$

3223 1) $^9/_{12}$ or $^3/_4$
2) (a) $^7/_{12}$
(b) $^{10}/_{12}$ or $^5/_6$
(c) $^{11}/_{12}$
(d) $^{10}/_{12}$ or $^5/_6$
(e) $^7/_{12}$
(f) $^8/_{12}$ or $^2/_3$
(g) $^8/_{12}$ or $^2/_3$
(h) $^9/_{12}$ or $^3/_4$

3323 1) ¼
2) ¼
3) 8

3324 2) 8″ × 8″ : 1
4″ × 8″ : ½
$2^2/_3$″ × 8″ : $^1/_3$
2″ × 8″ : ¼
1″ × 8″ : $^1/_8$
½″ × 8″ : $^1/_{16}$
$^1/_3$″ × 8″ : $^1/_{24}$
3) Areas (in square inches) : $2^2/_3$; 4; 8; 16; $21^1/_3$; 32; 64.

3333 2) 8 × 8 : 1
4 × 8 : ½
$2^2/_3$ × 8 : $^1/_3$
2 × 8 : ¼
1 × 8 : $^1/_8$
½ × 8 : $^1/_{16}$
$^1/_3$ × 8 : $^1/_{24}$
5) (decreasing) : ½, $^1/_3$, ¼, $^1/_6$

3343 2) (a) ¼
(b) $^1/_{12}$
(c) $^2/_6$ or $^1/_3$
(d) $^1/_{12}$
(e) $^2/_{12}$ or $^1/_6$
(f) $^3/_{12}$ or ¼

3353 2) (a) 6
(b) 2
(c) 4
(d) $^4/_3$
(e) $^1/_3$
(f) $^1/_8$

3713 Pi = $\pi \approx 3.14$

3723 Look up golden ratio or golden section.
$\frac{l}{w} \approx 1.6$

3733 Same ratio since units cancel.

3743 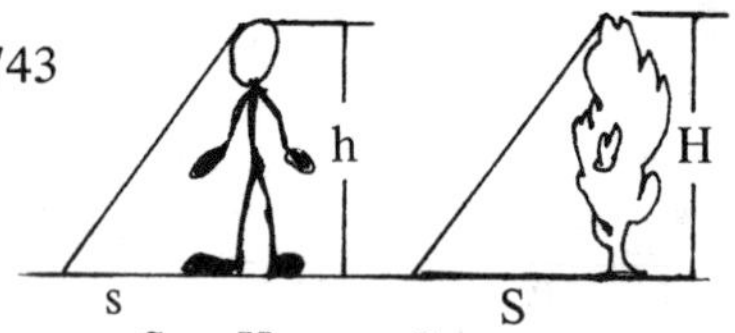

H – tree's height
h – your height
S – tree's shadow
s – your shadow

$\frac{S}{s} = \frac{H}{h}$ or $\frac{S}{s} \times h = H$

3753 Weight is proportional to volume, assuming that the model and the real thing have the same density. So the weight scale will be the cube of the linear scale.

3763 2) No, it is 8 times larger.
$^1/_8$
$^1/_{8000}$
$^1/_{64000}$

4124 P = $^2/_3$
Probability should be about $^2/_3$.

4154 Probability should be ½.
The coin cannot remember.

4173 Regard each of the possible numbers as being one less, 0 - 999. Then consider each as being made up of three digits, each digit being a possible choice of 0 - 9. These choices can be selected, in order, by spinning a spinner evenly spaced with numerals 0 - 9 on it. To arrive at the six prize - winning numbers, add one to the six numbers selected by the above procedure.

4214 1) (1) 7,000
(2) 2,500
(3) no

4343 1) 20; 16½; 15
2) 20; 20; 14

VOLUME 2

5543 In general we have:

which when opposite corner numbers are added, gives 2a + 8 = 2a + 8.

5553 2) No; there is a maximum of 31 days in any month.
3) 365 ÷ 7 = 52 R 1

5573 1) (a) 0° (h) 135°
(b) 90° (i) 165°
(c) 180° (j) 82°
(d) 90° (k) 97°
(e) 60° (l) 67°
(f) 105° (m) 7°
(g) 135° (n) 142°

5583

IN:	NOON WOULD BE:
Chicago	early morning
Honolulu	middle of the night
Tokyo	late in the evening
Bombay	late afternoon
Cairo	mid afternoon
Rome	early afternoon

5944 The volume of irregular shapes can be found by immersing them in a full container and measuring how much liquid overflows. Or if the container is rectangular or cylindrical, by filling it enough to cover the object, and calculating the volume by the rise in water level.

5954 2) Mercury; its density is 13.55 grams per cubic centimeter, or 13.55 times the density of water.

6123 Triangle, in construction.

6124 Rectangle, 4, opposite edges, 4, all corners the same.

6134 1) yes
2) yes; rectangles

6143 Parabola

6154 area "changes" from 64 to 65.

6163 area "changes".

6213 Triangle, 3.

6223 1) This depends on the number of nails.
2) With 12 nails - regular 3-gon, 4-gon, 6-gon, 12-gon
48 nails – as above, plus 8- , 16- , 24, 48-gon.
49 nails – 7- , 49-gon.

6233 The diagonals of any rectangle will intersect at the center of the circle.

6243 1) By the nail spacing.
2) Yes, but it will be a square.
3) Yes, but it will be a rectangle.

6253 1) Yes
2) They intersect at right angles, and one is a diameter.
3) One opposite pair are right angles, and the other are supplementary.

6263 No concave quadrilateral can be cyclic, because one of its angles must be reflex, and hence could not be supplementary to that opposite to it.

6314 Circle, infinite, circle, point.

6324 Pi (π)

6374 1) $\angle$ QTS = $\angle$ TRS
2) 90°

6413 1) yes
2) yes
3) yes
5) Research

6414 3)

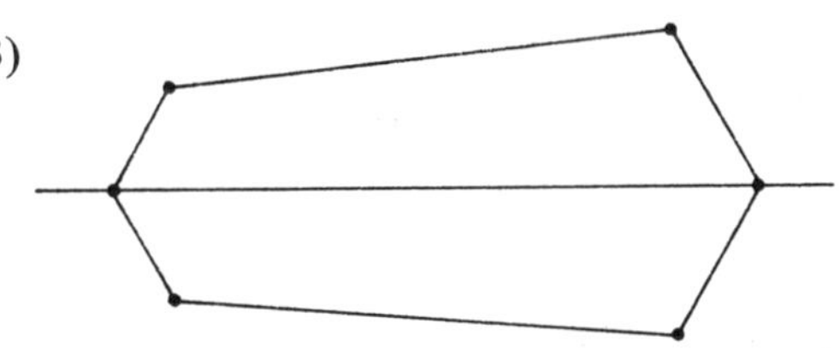

6424 1) 0, 8
2) 101, 202, 303, . . ., 111, 121, ...
3)

6434 5)

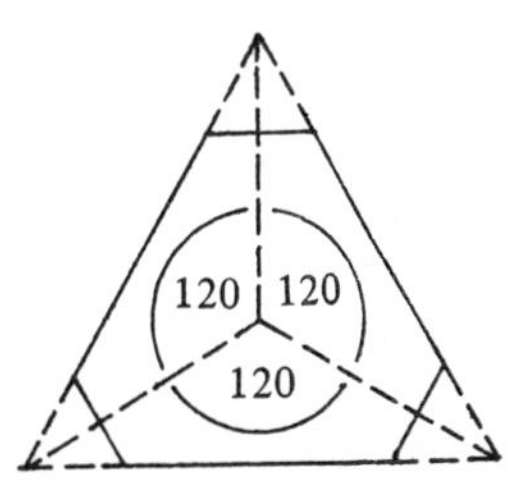

VOLUME 2

6444 1) True
2) Not always true
θ (theta) ϕ (phi) o (omicron) χ (chi) s (sigma)

6453 one, none, three

6463 two

6473 same as the number of folds.

6474 No. of Cut-out Parts: 1, 2, 4, 8, . . . , 2^{n-1}
No. of Axes of Symmetry remains 2 after the first fold.

6524 3, 6, 10, 15.

6814 Right triangle, right angle.

6834 Diagonals in order have lengths $\sqrt{2}, \sqrt{3}, \sqrt{4}, \sqrt{5}, \ldots$

6923 2) A cube has no more than three faces meeting at any vertex.

6933 2)

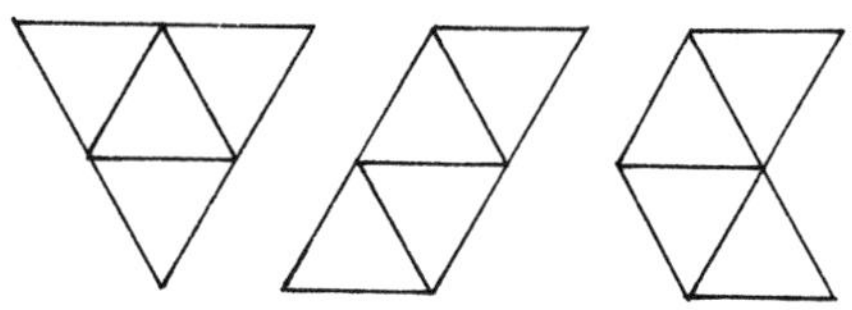

6953 2)

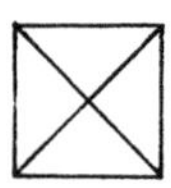

3)

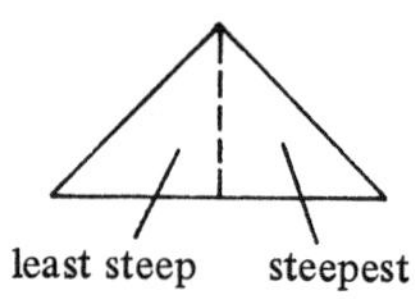

7113 No.

7244 3)

SIDE	AREA		
1	6	or	6.1^2
2	24	or	6.2^2
3	54	or	6.3^2
.	.		
.	.		
.	.		
n	$6n^2$		

7283 4) $2r^2 < \pi r^2 < 4r^2$

7294 Parallelogram, πr base and r height
So area = $\pi r \cdot r$ or πr^2.

7323 It weights less.

8113 A = {triangles and squares}
B = {squares, rectangles}
A and B = {squares}
A or B = {triangles, squares, rectangles}

8133

No. of Cans	No. of Moves
1	1
2	3
3	7
4	15
n	$2^n - 1$

8163 a) 6
b) 12
c) 8
d) none
e) one

8313 (a) 7 (b) 14 (c) 3 (e) 9 (f) 16
(g) 15 (h) 4 (i) 20

8323

13	8	15	10
12	5	7	12
11	3	8	11
10	8	15	13

8333

8	5	13	10
9	2	7	9
9	3	6	9
10	5	13	8

8343 1) 45
2) 15
5) 5; middle
6) 2, 4, 6, 8; corners

8363

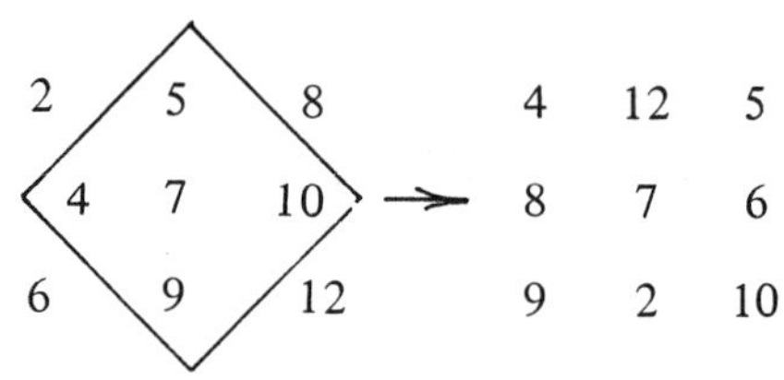

8373 1) (1) yes
(2) yes
2) no
3) yes

VOLUME 2

8383 1) Reverse the order of each diagonal
2) 4 rows, 4 columns, 2 diagonals, 4 quadrants, the middle quadrant, the 4 corners – total: 16
3) 111; 505

8393 It is possible.

8423 9, 6

8424 (1) 24, (2) 13

8433 1) 5
2) 14
3) 55

8443 1) B, C, D, E, H, I, K, O, X,
2) A, H, I, M, O, T, U, V, W, X, Y
3) horizontal
4) none

8473 a) 3
b) 4

8483 1) no
2) rhombus
3) yes
4) No. Base and height
5) yes

8493 1) 2 - parallelogram, kite
2) no
3) yes
4) triangle

8513 5) can't be done.

VOLUME 3

NOTE: *Experiments not listed* are experiments with student research answers. The student research answers will vary.

1155 c) Triangle

1175 Add and subtract.

1355 $2 \times 2 = 4$
6 towers: 2; 2×2; $2 \times 2 \times 2$, etc.
2^2, 2^4, 2^6 are perfect squares.

1525 $1 + 4 \cdot 2$
$5^2 = 3^2 + 4 \times 4$
$7^2 = 5^2 + 4 \times 6$
⋮

2115 Each letter corresponds to exactly one city.

2315 1) Use only 7 fingers per man.

2435 3) 3×5, 5^2, 3^3, $2 \times 3 \times 5$.

2555 1) No, $(41^2 - 41 + 41) \div 41$.

2565 a) $13 + 7$
b) $13 + 17$
c) $43 + 19$ or $31 + 31$
d) $53 + 29$

2635 2) Rule out any divisor in a line where it will not divide evenly into one or more of the numbers. In the above case, this will rule out all the divisors except the first 2. So the G.C.D. must be 2.

2815 2) Sum of the first n odd numbers is n^2.
3) 49, 121, 400.
4) $400 - 20$, or 380.

2825 1)

1	1
3	4
5	9
7	16
9	25
11	36
13	49
15	64
17	81
19	100

2) The square numbers. Successive differences give the odd numbers.
3) The nth number in the sequence is $2n - 1$, and this triangle encloses n^2 nails.
81, 400, 900.

2835 1)

1	1
2	3
3	6
4	10
5	15
6	21
7	28
8	36
9	45
10	55

2) 4; 9; 400; 10,000.
3) You get the right one.

2845 1)

1	1
2	5
3	13
4	25
5	41
6	61
7	85
8	113
9	145
10	181

2) $2n^2 - 2n + 1$;

15	421
20	761
100	19,801

2915 $3^2 + 4^2 = 5^2$
$5^2 + 12^2 = 13^2$
etc.

3145 Four times as large.

3415 $^1/_7 = .142857\ldots$
$^2/_7 = .285714\ldots$
$^3/_7 = .428571\ldots$
$^4/_7 = .571428\ldots$
$^5/_7 = .714285\ldots$
$^6/_7 = .857142\ldots$

VOLUME 3

3725 2) no
3) yes

4135 4.5

4145 4.9

4196 1) 24

4336 1) should be 50%.
3) should be close to 90%.

4346 3) Favors the side of Lincoln's head.

4365 Pi = $\pi \approx 3.14$

4375 10.6

4385 9.6

5135 about 6.28 inches.

5145 only the length of the radius.

5215 area 4 times.
volume 8 times.

5225

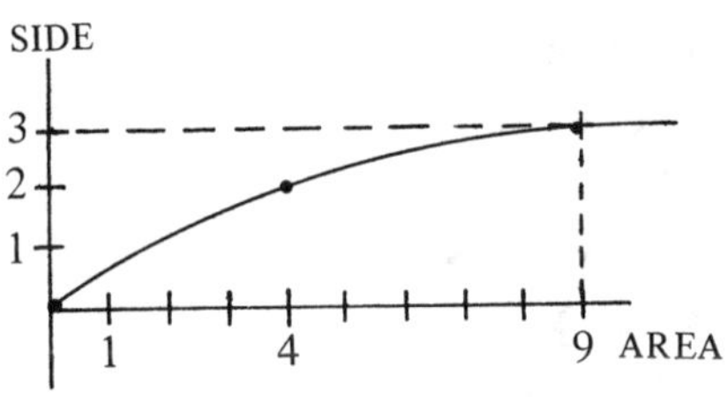

5325 cord: cut wood equal to a stack 4 × 4 × 8 feet.
board foot: quantity of lumber equal to a board 12 × 12 × 1 inch.

5355 4) 0.002183
0.000157
0.032580
0.0000023
537,000,000
22,900,000
70,000

5525 1) 165
247½
330

2) By interpolation, $12{:}32^{8}/_{11}$; $12{:}16^{4}/_{11}$; $12{:}49^{1}/_{11}$; $1{:}05^{5}/_{11}$.

5535 Yes; yes; the graph will be a straight line.

6126

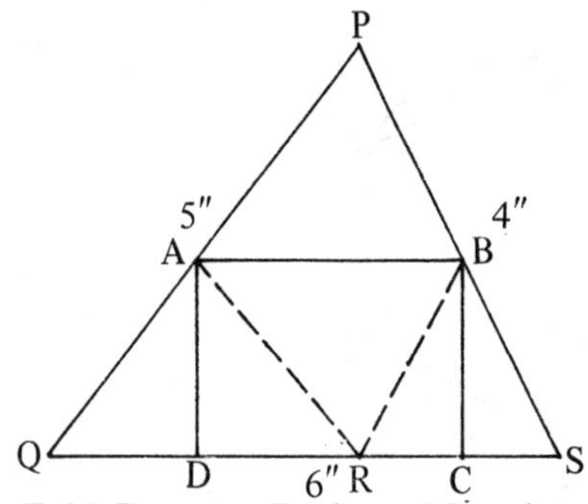

Fold P on to R along AB, where A and B are mid-points of PQ and PS. Then Q and S on to R. This results in a rectangle with the paper doubled, and no overlap.

6135 1) 4; infinity; 1.

2) One and two folds result in a rectangle twice as long as it is wide, or in a 45/45/90 triangle. Four folds produce a square half the area of the original.

6155

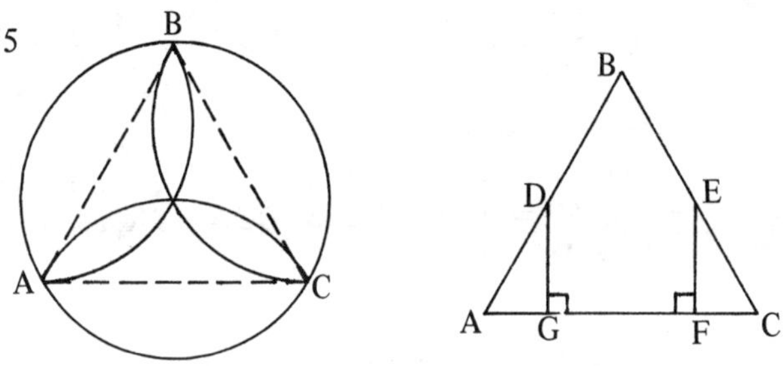

1) Fold the edges along AB, BC, CA in turn so that the edges coincide with the center of the circle.

2) Fold B on to G so that BD lies on DG. Select point E likewise. By symmetry DE is parallel to AC. Check that DG and EF are perpendicular to AC. Then DG = BD = DE.

6165

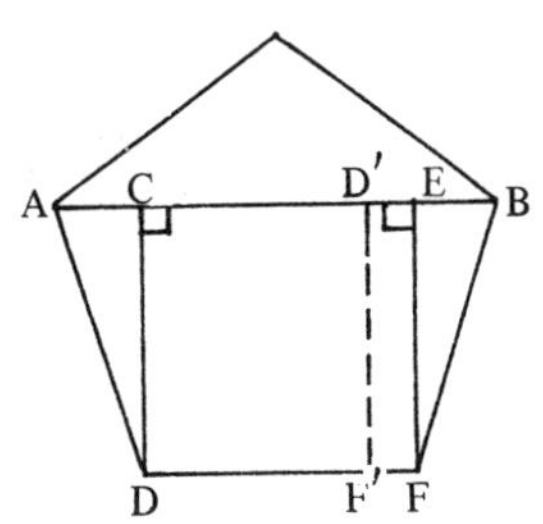

1) 5 ways, and each results in a general quadrilateral.
2) Each fold is on an axis of symmetry.
3) Folding along AB gives a trapezoid. Then along CD and EF gives a rectangle. Finally if D is folded to lie on CE, say at D′, and if F′ is similarly found, the rectangle CEFD can be folded along D′F′ to give a square CD′F′D.

VOLUME 3

6175

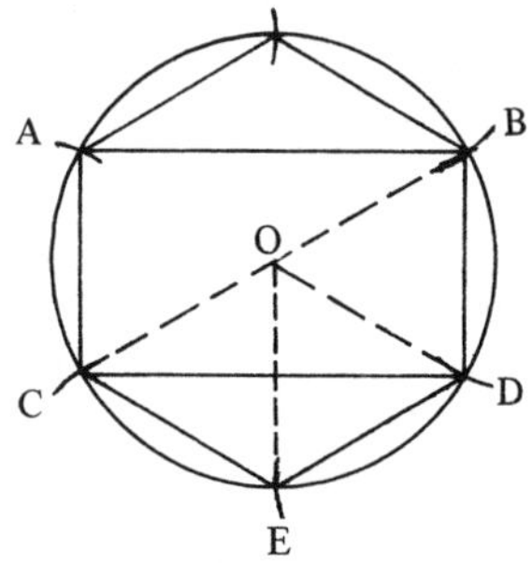

1) Fold along each possible axis of symmetry.
2) Folding along AB and CD gives a rectangle. Unfold and fold along BC to get an isosceles trapezoid. Then fold CE on to ED to get the rhombus OBDE.
3) Finally fold B on to E to get the equilateral triangle OED.

6185

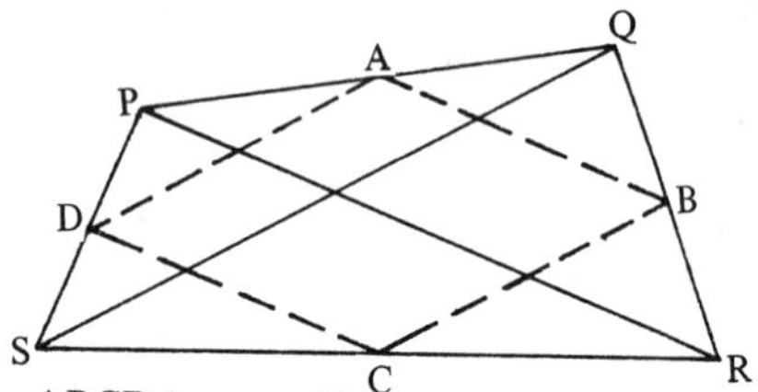

ABCD is a parallelogram. By data, AQ = ½ PQ, and thus AB = ½ PR. Likewise DC = ½ PR, and hence AB = DC. Similarly DA = CB.

6195 Let BC be parallel to DE so that area of Δ ABC = ½ area Δ ADE.

By similar triangles. $\frac{BP}{DQ} = \frac{AP}{AQ}$

By construction, 2 BP AP = DQ AQ

Therefore $\frac{BP\ AP}{DQ\ AQ} = \frac{1}{2}$

Substituting, we get $\left(\frac{AP}{AQ}\right)^2 = \frac{1}{2}$

Therefore AP : AQ = 1 : $\sqrt{2}$.

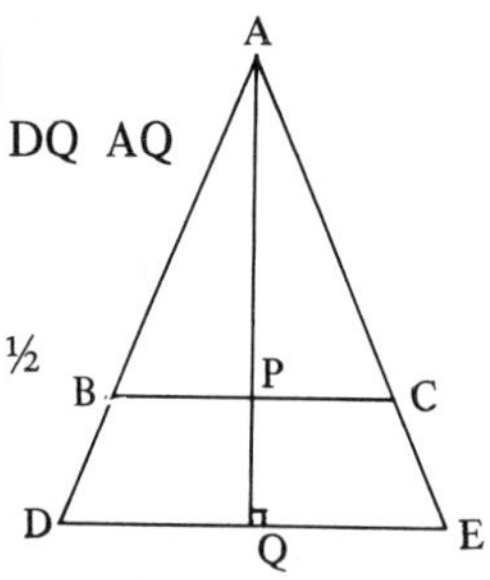

6215

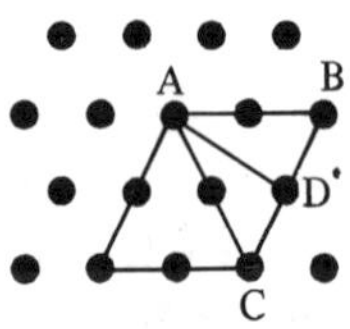

Δ ABC is equilateral. D is the mid-point of BC by symmetry. Therefore ∠ ADC is a right angle.

6225

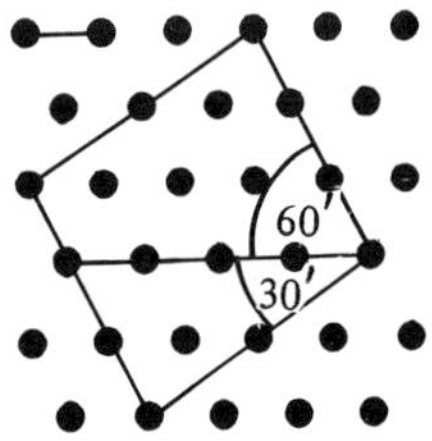

The top left shape is a rectangle, whose sides are 3a and $2\sqrt{3}$ a. The other shapes are rhombuses. It is not possible to make squares on an isometric geoboard.

6245

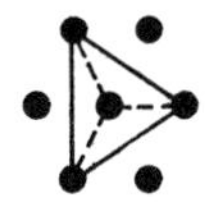

6255

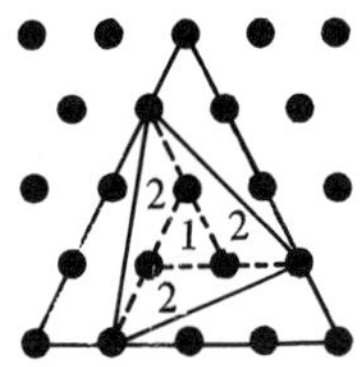

1) Yes, by symmetry
2) Small – 7 units
   Large – 16.

6265

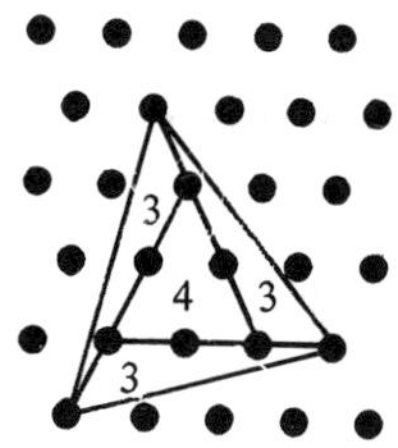

13

6275 1) 30°, 60°, 90°.
2) They are equal.
3) Yes.
4) No, the ratios are trigonometric ratios.

6285 1) $\frac{1}{4}\sqrt{3}$.
2) $100 \cdot \frac{1}{4}\sqrt{3}$ (approx. 43.3);
   $400 \cdot \frac{1}{4}\sqrt{3}$ (approx. 173.2)
   $10{,}000 \cdot \frac{1}{4}\sqrt{3}$ (approx. 433)

6295 Both quadrilaterals would be parallelograms.

6365 1) The diameter bisects the chord.
2) Diameter, semi-circles.

6366 1) ∠ AOB = 2 ∠ APB
3) ∠ APB = 90°

VOLUME 3

6375

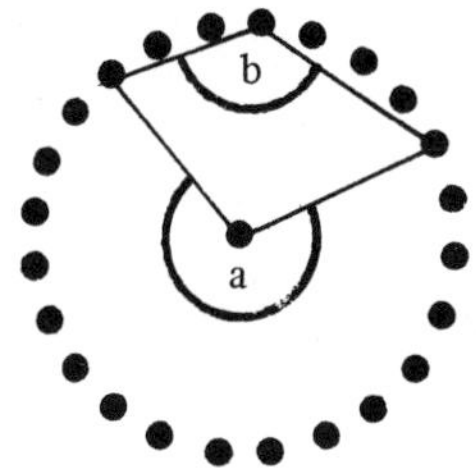

1) a = 2b (angle at the center). But in a rectangle, a = 3b.
2) The parallelogram would be a rhombus as two adjacent sides are equal. It must be a 60° rhombus, since **a** would be 240°, and **b** 120°.

6386 1) Parabola
2) Ellipse
3) Hyperbola

6425

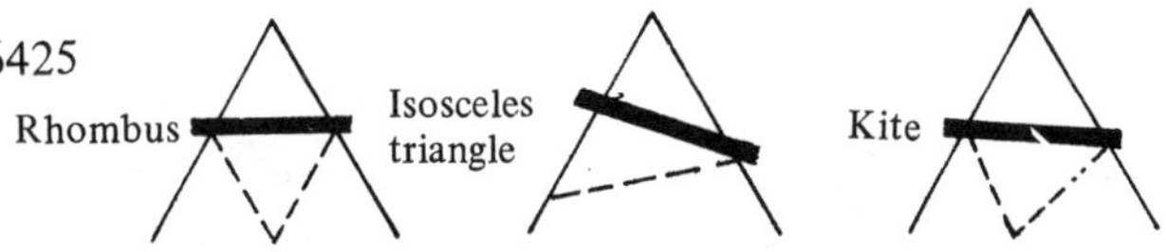

6435

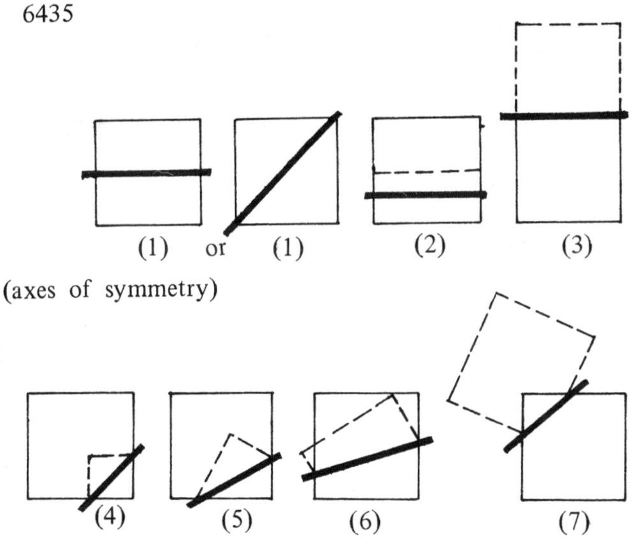

6445

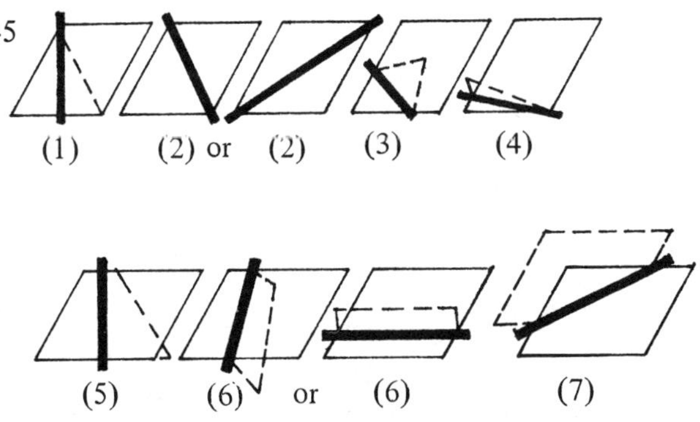

6576

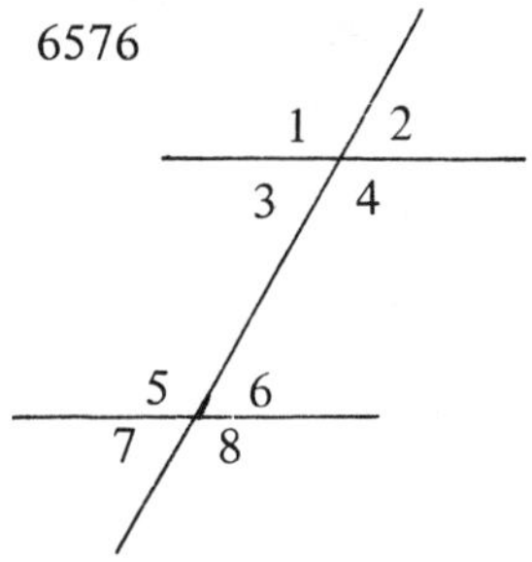

a) $\angle 1$ & $\angle 4$; $\angle 2$ & $\angle 3$
$\angle 5$ & $\angle 8$; $\angle 6$ & $\angle 8$
b) $\angle 1$ & $\angle 2$, $\angle 3$ & $\angle 4$, etc.
c) $\angle 1$ & $\angle 3$; $\angle 1$ & $\angle 2$, etc.
d) $\angle 1$ & $\angle 5$; $\angle 3$ & $\angle 7$, etc.
e) $\angle 4$ & $\angle 5$; $\angle 3$ & $\angle 6$
f) $\angle 1$ & 8; $\angle 2$ & $\angle 7$

$\angle 1 = \angle 4 = \angle 5 = \angle 8$
$\angle 2 = \angle 3 = \angle 6 = \angle 7$

6605 2) The lower right triangle has area 1; the other 1½.

6615 Each has area 2, but the lower-right triangle has no right angles.

6616 1) Yes
2) Yes (see diagram)

6625 1) No. The lower one has a side of 2 ins, and the other triangle has no 2 inch side.
2) They are isosceles - - each has a pair of sides, each of which is a diagonal of a 3 X 1 rectangle.

6626 1) No. The top left has a side of 2 ins, but neither of the other triangles has. Of the two triangles on the right, the lower one has a side of 3 ins, and the other has not. Hence none of the three triangles is congruent.
2) No. The sides of a skewed right angle are formed only from the diagonals of congruent or similar rectangles. The only "possible" right angle has as its sides the diagonals of a 2 X 1 and a 3 X 1 rectangle, which rectangles are not similar.
3) The lower left angle of the lower triangle is 45°, formed by the side and diagonal of a 2 X 2 square.

6635 1) The opposite sides are, in each case, diagonals of 2 X 2 and 1 X 1 squares.
2) Diagonals of squares are proportional to their sides. So the ratio is 2:1.

6636 1) Kites
2) Yes. Corresponding sides and angles are congruent.

6645 1) Trapezoids.
2) No. The lower one has the shorter of the parallel sides 2 ins. The corresponding side in the other trapezoid is not 2 ins. Neither is an isosceles trapezoid.

6646 1) Parallelograms.
2) No. While each has the shorter pair of sides congruent, the longer pair are not congruent.

VOLUME 3

6655 1) Parallelogram, rhombus.
2) No. The top quadrilateral is not equilateral, while the other one is.

6665 1) No.
2) No. The top quadrilateral has a side of 3 ins., but the lower one does not; the lower one has a side of 1 inch, but the top one does not.
3) No.
4) Yes, the angles must also be equal, each to each, for congruence.

6675

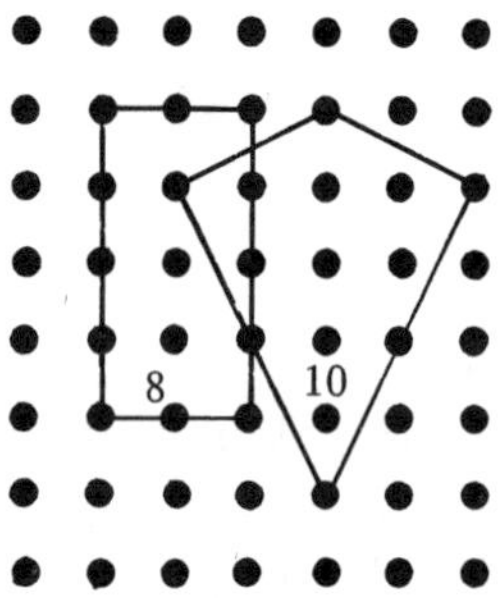

6685 1) Yes. Their corresponding sides and angles are congruent.
2) As above.

6695

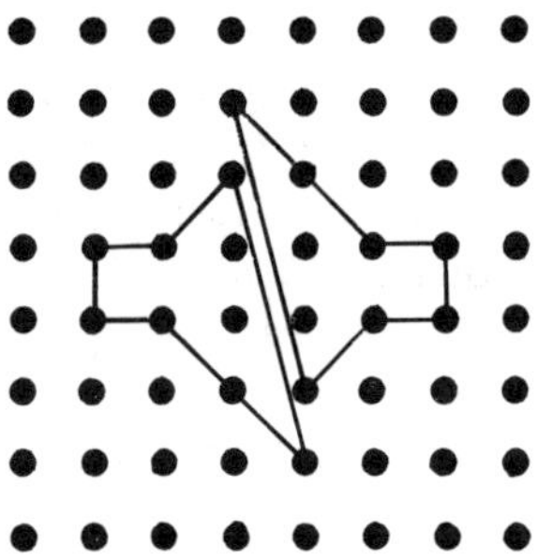

6705 1) The top two. They have corresponding sides congruent.
2) The top tow, because congruent figures must also be similar (though not necessarily vicea versa).

6715 1) Similar figures have their corresponding angles congruent.
2) Each is a right triangle whose shorter sides are in the ratio 2:1.

6735 1) No. Their areas are 3½, 5 and 4.
2) The top two triangles. Each is a right triangle with its shorter sides in the ratio 1 : 2.

6745 1) All are similar. The ratio of adjacent sides in each is 2 : 1.
2) Circles, and all regular polygons.

6755 Yes. Corresponding sides are in the ratio 1 : $\sqrt{2}$. Corresponding angles can be shown by trigonometry to be congruent.

6765

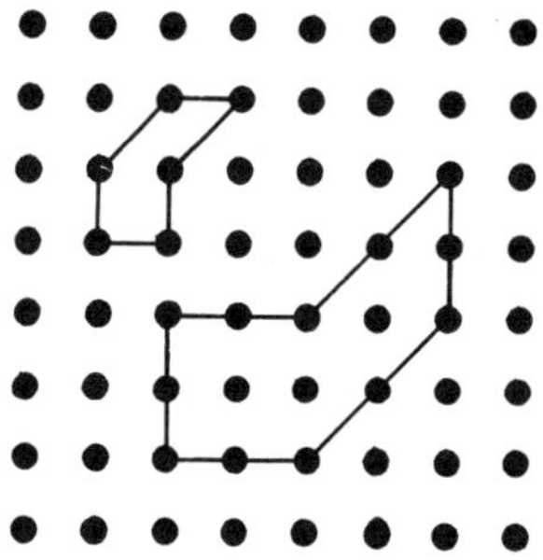

Areas are 2 and 8.

6775 1 inch and 2 inches.

6785 1) 1, 2, 3, 4 inches.
2) Yes.
3) 1¼, 2½, 3¾ inches.

6805 The sum of the two smaller is equal to the larger.

6835 Top Left: 1, 16, 17.
Top Right: 1, 9, 10.
Middle: 5, 5, 10.
Lower: 5, 20, 25.

6845 No. The two larger squares are congruent.

6855 1) Top: 2, 5, 9.
Lower Left: 4, 5, 13.
Lower Right: 5, 8, 9.
There is no relationship.

6865 $\sqrt{26}$; $\sqrt{2}$; $\sqrt{5}$; $\sqrt{10}$; $\sqrt{17}$.

6875 3) 5, 13.

6915 Pyramid.

6925 4) Five.

6935 Circle, ellipse or parabola.

VOLUME 3

6945 $V + F = E + 2$

Name	Vertices	Faces	Edges
		Number of	
Tetrahedron	4	4	6
Octahedron	6	8	12
Square pyramid	5	5	8
Hexagonal prism	12	8	18

6955 1) Cube, cuboid.

7115 1) Yes, e.g., 2 X 2 and 4 X 1.
3) There is no relationship.

7125 1) Yes, e.g. 3 X 2 and 4 X 1.
2) Yes.

7155 1) Parabola.
2) Circle, area approx. 103 sq. in.

7206 None are congruent nor similar. The top left is isosceles, and the lower left is not. Two have area 2, and the other two have area 3.

7215 1) 3¾.

7216 1) Yes. ∠b + ∠c = 90°. a = c (congruent triangles) Therefore, a + b = 90°.
2) The two shorter sides are both diagonals of 3 X 1 rectangles.
3) 5.
4) The square of which this triangle forms one-half has area 10. Therefore the length of the shorter sides is $\sqrt{10}$.

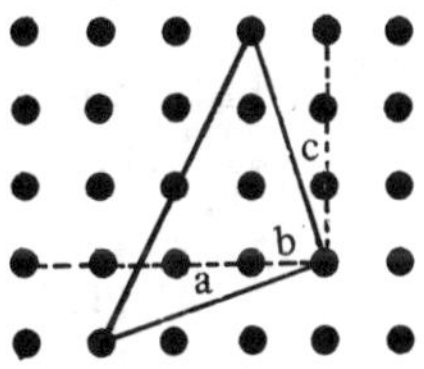

7226 1) Each has area 1.
2) Equal bases and heights.
3) ½•base•height.

7235 1) 3 and 2½.

7236 1) base 1, height 5.
2) 2½
3) Each is another triangle of area 2½. 5 + 5 = 10. 10 X 2½ = 25, which we know is the area altogether.

7245 2) 3
3
10
$4^3/_8$
$12^3/_8$

7255

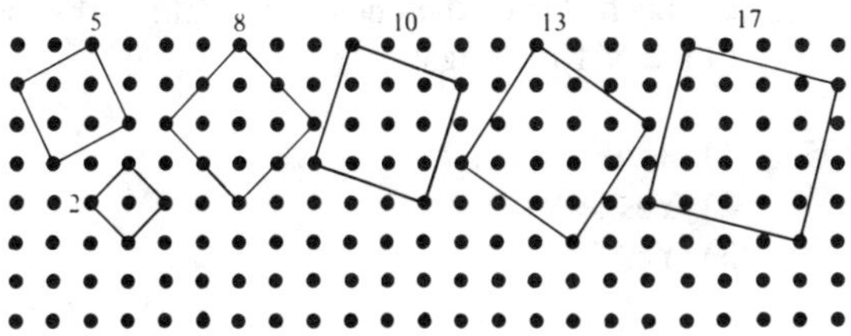

7256 1) True

7265

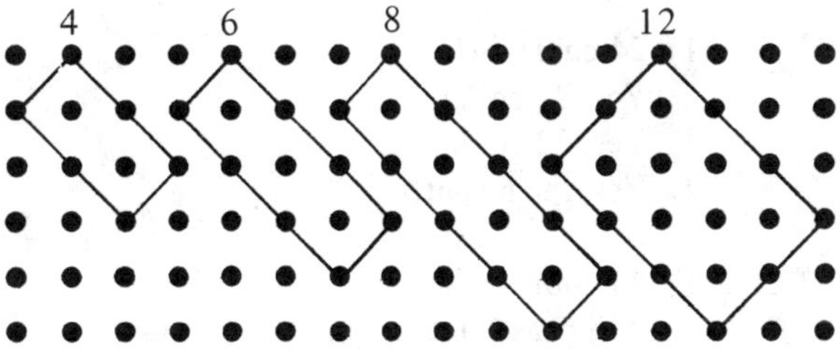

7266 1) Base 3, height 1. Perimeters not equal.
2) Each area is 3.
3) Base • height.

7276 1) Parallelogram of base 4 and height 4. Its slant height must be greater than 4. Area is 16.
2)

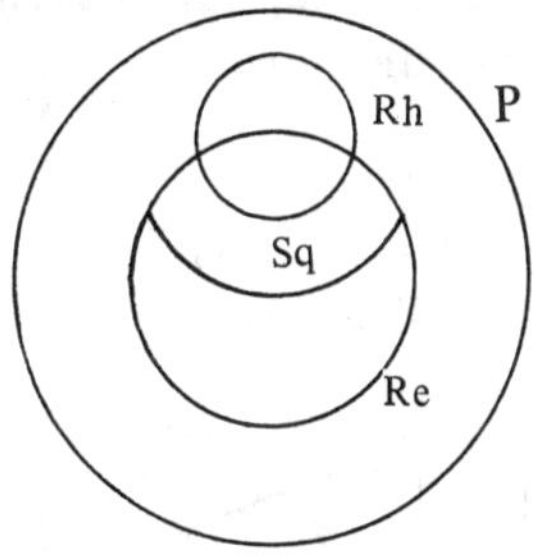

7285 1)

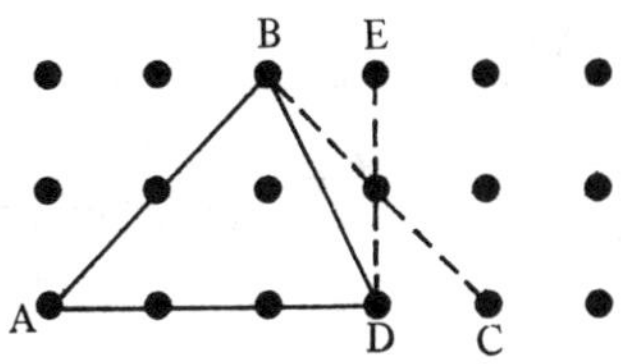

∠BAD is 45° (diagonal of square).
∠ABD is less than ∠ABC, which is a right angle.
∠ADB is less than ∠ADE, which is a right angle. Hence the triangle has no right angles.

2)

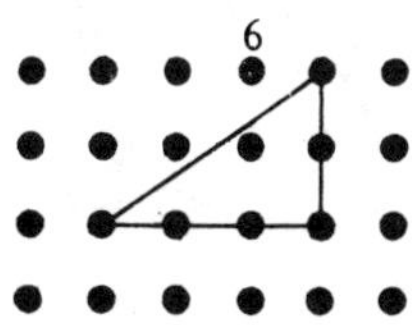

VOLUME 3

7286 1) Areas are 5 and 4.
2) Each side of the one on the right is the diagonal of a 2 X 1 rectangle.

7295 1) 3
2) Yes
3) Yes

7296 1) Areas are 4 and 6.
2) Each diagonal of a rhombus divides it into two isosceles triangles, and no isosceles triangle can have base angles right angles.

7315 1) 24 cubic inches.
2) Yes, 23 cu. in.
3) Yes, 22 cu. in.
4) Yes; the height.

7325 1) 44 cubic inches
2 in.; base area
2) 3•a; 4•a; h•a
3) height; base area

7326 1) Rectangle, approx. 35 sq. in.
2) 11 in.; 6 paths
3) Tetrahedron, approx. 21 sq. in.

7345 The volume is the same. They have the same length and the same cross-section area.

8125 5 Days = 2^5 = 32 cents
6 Days = 2^6 = 2 (32) = 64 cents
20 Days = 2^{20} = 1,048,576 cents

8415

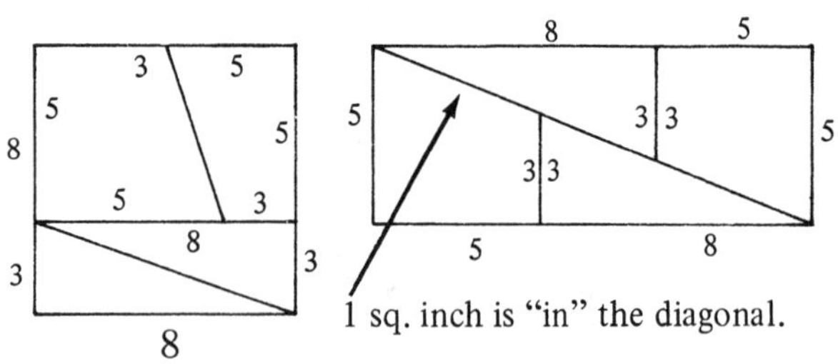

8425 2) 2
3) Yes, a trapezoid

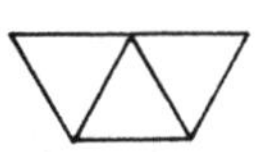

4) Yes
5) No, only the square and pentagon.

8455 2) Yes
3) Yes; no

8465 2) No
3) No

8515 3) The ratio, diagonal of a regular pentagon: its side, is the golden ratio.
4) The U.S. flag.

8516

	Regions	**Vertices**	**Edges**
A	2	3	3
B	2	4	4
C	1	2	1
D	1	4	3
E	3	4	5

R + V = E + 2

8525 List A is polygons of side 3, 4, 5, 6, and 7.
List B is a hexagon with all possible diagonals drawn from any one vertex to other vertices in succession.
List C is a heptagon - ditto.

8535 List D is a quadrilateral with 0, 1, and 2 diagonals.
List E is a pentagon with all possible diagonals drawn, working clockwise, and moving from vertex to vertex clockwise.
List B completes all possible diagonals in a hexagon in the same manner as for the pentagon.

Regions	Vertices	Edges
14	12	24
17	14	29
22	18	38
26	21	45

List C completes all possible diagonals in a heptagon as above

Regions	Vertices	Edges
11	10	19
15	13	26
20	17	35
23	19	40
28	23	49
35	29	52
39	32	59
46	38	72
51	42	81

General notes: Start at any vertex, and draw diagonals in turn, working clockwise. Then move clockwise to the next vertex, and repeat. Ensure that no more than two diagonals intersect at any point.
In general, when a new diagonal crosses n regions, it forms an additional n regions, and an additional (n − 1) vertices. So if the first column goes up by 3, the second would go up by 2. This would form 5 new edges.

VOLUME 3

8545 1) The number of trios is the sequence of triangular numbers, thus:

n-gon	trios
3	1
4	3
5	6
6	10
7	15

The list beginning 2 8 8 will therefore have 21 trios.

2) 6 kites, 12 right triangles, 6 isosceles triangles. In the regular hexagon, three lines meet at the center, obscuring what would, in a non-regular hexagon, be the 26th region (a triangle).

8555 1) Yes
2) In the heptagon; yes.

8565 The heptagon, hexagon, and pentagon only need two colors.

8575 G: 1, 3, 2
H: 1, 4, 3
I: 1, 5, 4
n = 2

8585 K: 4, 4, 6
L: 8, 4, 10

8595

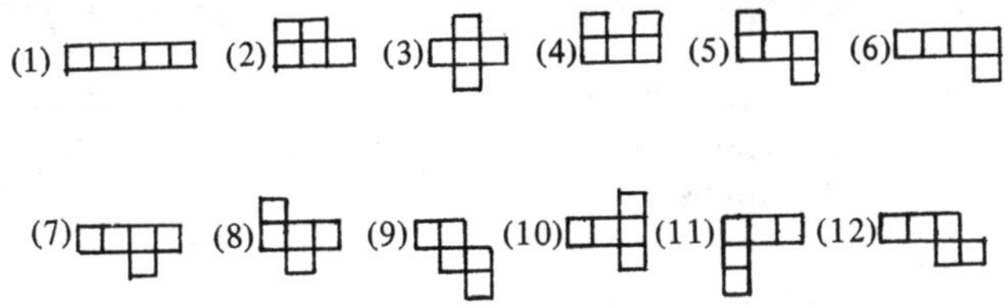

8725 One solution:

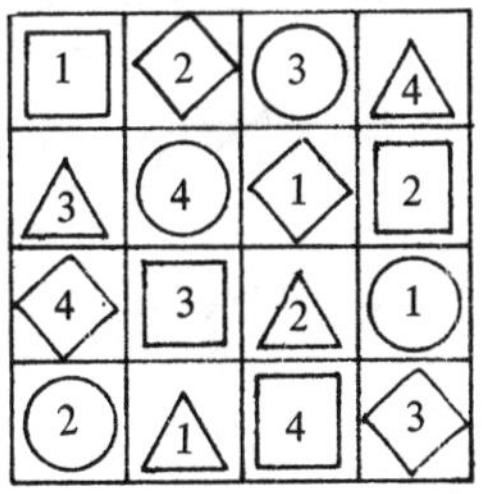

(Numbers refer to colors.)

8755 1)

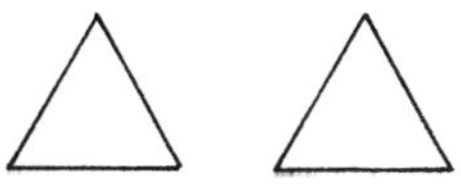

2)

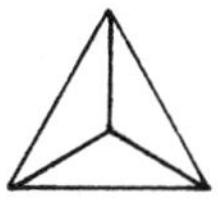

3)

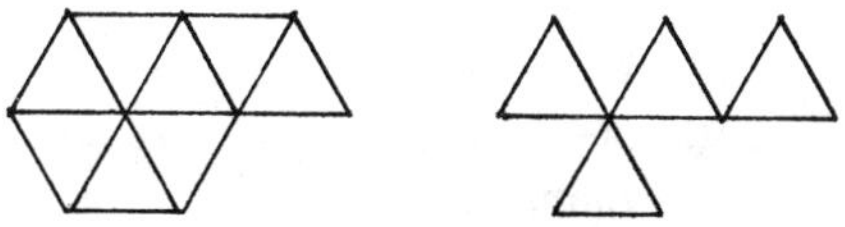

8765 1)

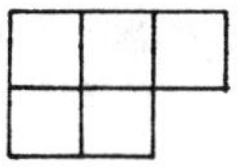

2)

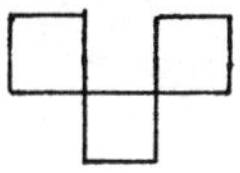

3)

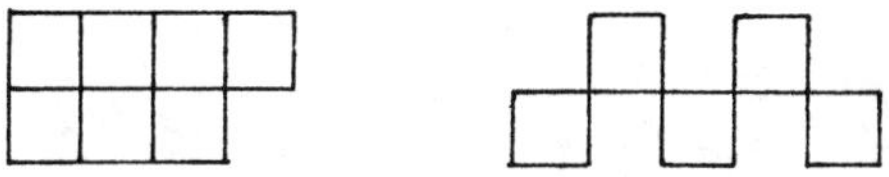

4)

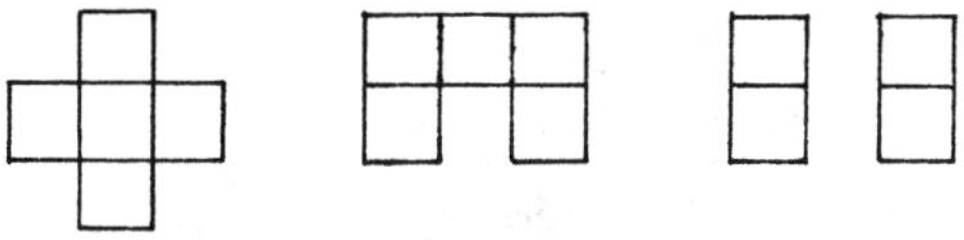

VOLUME 4

NOTE: *Experiments not listed* are experiments with student research answers. The student research answers will vary.

1317 2^{50} sheets.

2517 In the second century you must also rule out multiples of 11, 13, and 17 – namely 121, 143, 169, and 187, making 21 primes in the second century.

2537 The number of primes in successive centuries does gradually get less, because the number of prime factors available to render the numbers composite increases. But the reduction in numbers of primes in successive centuries is not regular or constant, nor do you ever reach a final prime.

2546 4) They were removed with the multiples of 2.
5) Because all multiples of 4 also are multiples of 2.
7) They are all the primes in the range, 1 - 100.

2567

x	f (x)	g (x)
0	41	1601
1	41	1523
2	43	1447
3	47	1373
4	53	1301
5	61	1231
6	71	1163
.		
.		
.		

One observation: Tabulated function values are primes.

2817 36.
Each red shows factor of 2; each green shows factor of 3. Perfect squares have an *even* number of reds and an *even* number of greens.
Perfect cubes have 3n reds and 3p greens.

2827 1) 23,

2)

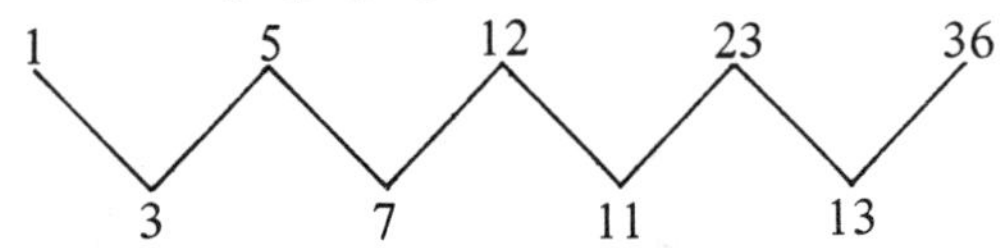

2837

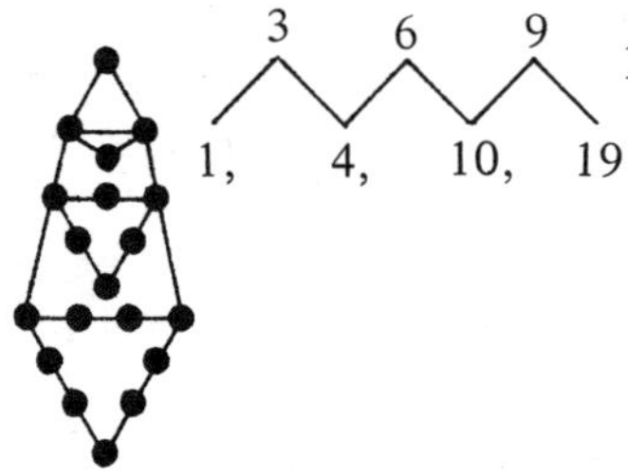

1 + 4 + 10 + 19 = 34

2847

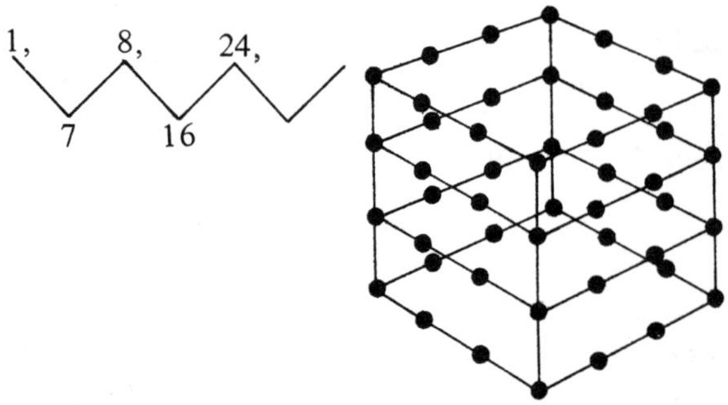

2857 Perfect: sum of proper divisors is the number.
Abundant: sum of proper divisors is greater than the number.
Deficient: sum of proper divisors is less than the number.

2917 3) Example: 36 – 1 – 3 – 5 – 7 – 9 – 11 = 0
So: $\sqrt{36} = 6$

2957 1) 5; 5
2) 9
3) 400; 201; $(1 + 400) \bullet \frac{200}{2}$

4117 Not a fair coin, probabiltiy will not by 50% for heads or tails.

4127 Should be close to 32 heads or a total of 32.

4137 2) no.

4157 Almost 100% of the time.

4167 Pi = $\pi \approx 3.14$

4197 a) 36
b) one
d) about $\frac{1}{6}$

VOLUME 4

4236 Normal curve.

4237 Should be close to the starting place.

4246 Normal curve.

4247 Should be close to 50% heads.

4357 Close to 50% for 23 students.

5117 Should be close to the starting point.

5327 4) 1.276
0.283
0.0153
0.0017
137,000
2,750
80

5717 A – 440

6117 2; 8

6217 1) No; no
2) In a trapezoid, two pairs of adjacent angles are supplementary. So if one angle is a right angle, another must be as well. If three angles were right angles, the fourth would be as well (angle total 360°). The figure would then be a rectangle.

6227 A polygon all of whose sides and angles are congruent.
True.

6237 Orthocenter, incenter, circumcenter, centroid.

6258 180°

6267 360°, 540°

6278

No. of sides	No. of diagonals one vertex	No. of total diagonals
3	0	0
4	1	2
5	2	5
6	3	9
7	4	14
8	5	20
10	7	35
n sides	n - 3	$\frac{n(n-3)}{2}$

6288 Sum should be equal to 180°.

6297 Number of sides, No. of Diagonals, No. of triangles

No. of sides	No. of diagonals (one vertex)	No. of triangles	Sum of angles
3	0	1	180°
4	1	2	360°
5	2	3	540°
n	n-3	n-2	(n-2) (180°)

6397 2) ellipse
3) Sum should always be equal.

6487 a) one b) many c) many d) one e) none
f) one g) none h) none i) three j) one

6497 1) many 2) two 3) five 4) three

7207 1) A kite is a quadrilateral with two pairs of adjacent sides equal.
2) See diagrams below.

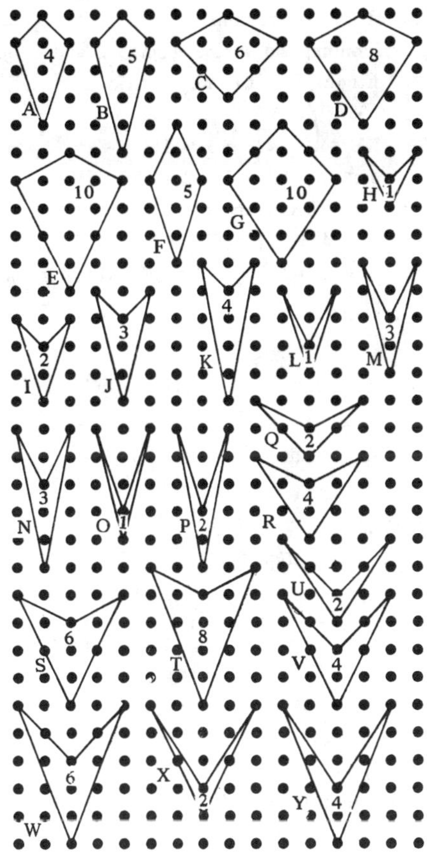

3) Shown above. The area of a kite is one half the product of its diagonals.
4) Kites A, B, G, Q have one right angle. E has two. All except A through G have reflex angles.
5) A kite's diagonals intersect at right angles, and one is bisected by the other.

VOLUME 4

7218 6½

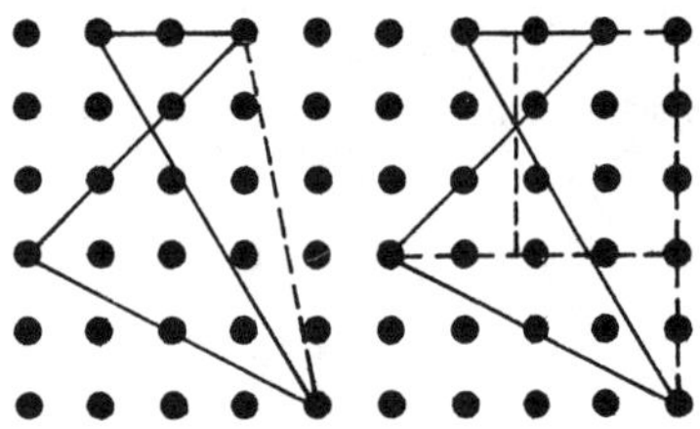

In fig. 1: $a + b = 20 - 4 - 4\frac{1}{2} - 2\frac{1}{2} = 9$

$a + c = 15 - 7\frac{1}{2} - 2\frac{1}{2} = 5$

Subtracting: $(a + b) - (a + c) = 9 - 5 = 4 = b - c$

In fig. 2, area $\triangle ABH = 7\frac{1}{2}$

area $\triangle DCH = 7\frac{1}{2}\left(\frac{2}{5}\right)^2$

$= 1\frac{1}{5}$

(ratio of areas of similar △s equals the ratio of the squares of corresponding sides)

Area $\triangle$ GCH = 4

by subtraction, area $\triangle$ GDH = $2\frac{4}{5}$

base GD of $\triangle$ GDH = $2\frac{4}{5}$ (base of triangle = twice area divided by height)

△s AJE, GED are similar

Corresponding sides have lengths 2 and $2\frac{4}{5}$

heights KE, EL are in ratio $2 : 2\frac{4}{5}$

KL = 3

$KE = \left(\frac{2}{4\frac{4}{5}}\right) \cdot 3 = 1\frac{1}{4}$

area $\triangle$ AEJ = ½ 2 KE = $1\frac{1}{4}$ = c (fig. 1)

b − c = 4 (from above)

$b = 4 + 1\frac{1}{4} = 5\frac{1}{4}$

$b + c = 5\frac{1}{4} + 1\frac{1}{4} = 6\frac{1}{2}$, which is the required area.

7227 3) ½ • height • sum of parallel sides.

7237 1) Each has area 2.

7247 1) Yes

2) Yes; the areas are 1½, 3, 4½.

7257 1) 22½

7258 2) Yes; yes

(b) is directly proportional to (a)

(c) is not directly proportional to (a)

7267 1) By 2

2) If the area is ½•h (a + b), and you increase a by 1, the area will increase ½•h, which in this case is 2.

7268 1) Shape is immaterial.

2) Not all acres have the same perimeter.

7277

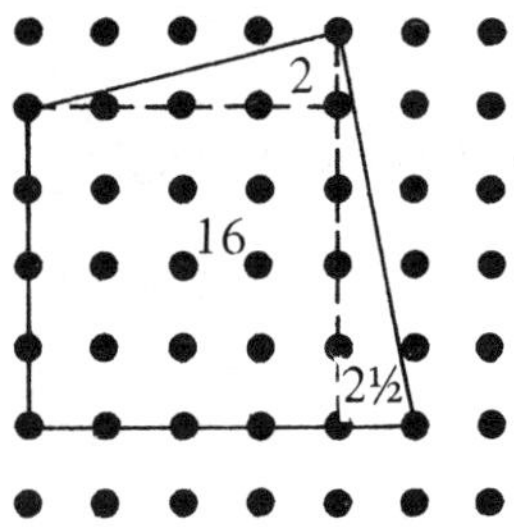

7287 Skewed squares have areas of 2, 5, 8, 10, 13, and 17.

7297

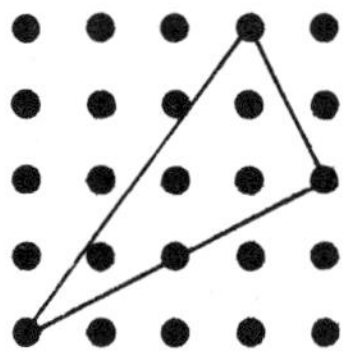

7327 Volume box 3 times volume of pyramid.

7337 Volume of cylinder 3 times volume of cone.

8487

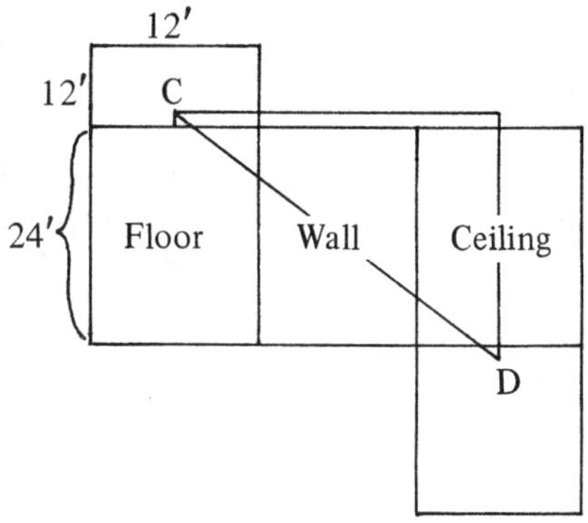

a) 40 ft.